BUGS

LADYBUGS

by Martha London

Consultant: Beth Gambro
Reading Specialist, Yorkville, Illinois

BEARPORT PUBLISHING

Minneapolis, Minnesota

Teaching Tips

Before Reading

- Look at the cover of the book. Discuss the picture and the title.
- Ask readers to brainstorm a list of what they already know about ladybugs. What can they expect to see in the book?
- Go on a picture walk, looking through the pictures to discuss vocabulary and make predictions about the text.

During Reading

- Read for purpose. As they are reading, encourage readers to think about the ladybug's life and the impacts the bug has on other things.
- If readers encounter an unknown word, ask them to look at the sounds in the word. Then, ask them to look at the rest of the page. Are there any clues to help them understand?

After Reading

- Encourage readers to pick a buddy and reread the book together.
- Ask readers to name three things from the book that ladybugs do. Go back and find the pages that tell about these things.
- Ask readers to write or draw something that they learned about ladybugs.

Credits:
Cover and title page, © Visual_Intermezzo/iStock; 3, © irin-k/Shutterstock; 5, © Chepko/iStock; 6, © Somchai Som/Shutterstock, © irin-k/Shutterstock; 7, © stoupa/iStock; 8–9, © Visual Intermezzo/Shutterstock; 10–11, © InsectWorld/Shutterstock; 12–13, © Ann in the uk/Shutterstock; 14–15, © Yevhenii Chulovskyi/Shutterstock; 17, © Jonathan Austin Daniels/iStock; 18–19, © Marko Zamurovic/Shutterstock; 21, © jorgen_norgaard/iStock; 22, © Andi111/Shutterstock, © HelloRF Zcool/Shutterstock, © KSCHiLI/Shutterstock, © mehmetkrc/Shutterstock; 23TL, © Darkdiamond67/Shutterstock; 23TR, © Scott Lyons/Shutterstock; 23BL, © Grant Terry/Shutterstock; 23BR, © gerasimov_foto_174/Shutterstock.

Library of Congress Cataloging-in-Publication Data

Names: London, Martha, author.
Title: Ladybugs / Martha London.
Description: Bearcub books. | Minneapolis, Minnesota : Bearport Publishing Company, [2022] | Series: Bugs | Includes bibliographical references and index.
Identifiers: LCCN 2021026698 (print) | LCCN 2021026699 (ebook) | ISBN 9781636913773 (library binding) | ISBN 9781636913841 (paperback) | ISBN 9781636913919 (ebook)
Subjects: LCSH: Ladybugs--Juvenile literature.
Classification: LCC QL596.C65 L66 2022 (print) | LCC QL596.C65 (ebook) | DDC 595.76/9--dc23
LC record available at https://lccn.loc.gov/2021026698
LC ebook record available at https://lccn.loc.gov/2021026699

For more information, write to Bearport Publishing, 5357 Penn Avenue South, Minneapolis, MN 55419. Printed in the United States of America.

Contents

Red and Spotted

A red bug lands on a leaf.

Black spots cover its body.

Soon, the **bright** ladybug will fly off again.

Ladybugs live all around the world.

There are many types of ladybugs.

But they are all smaller than a dime.

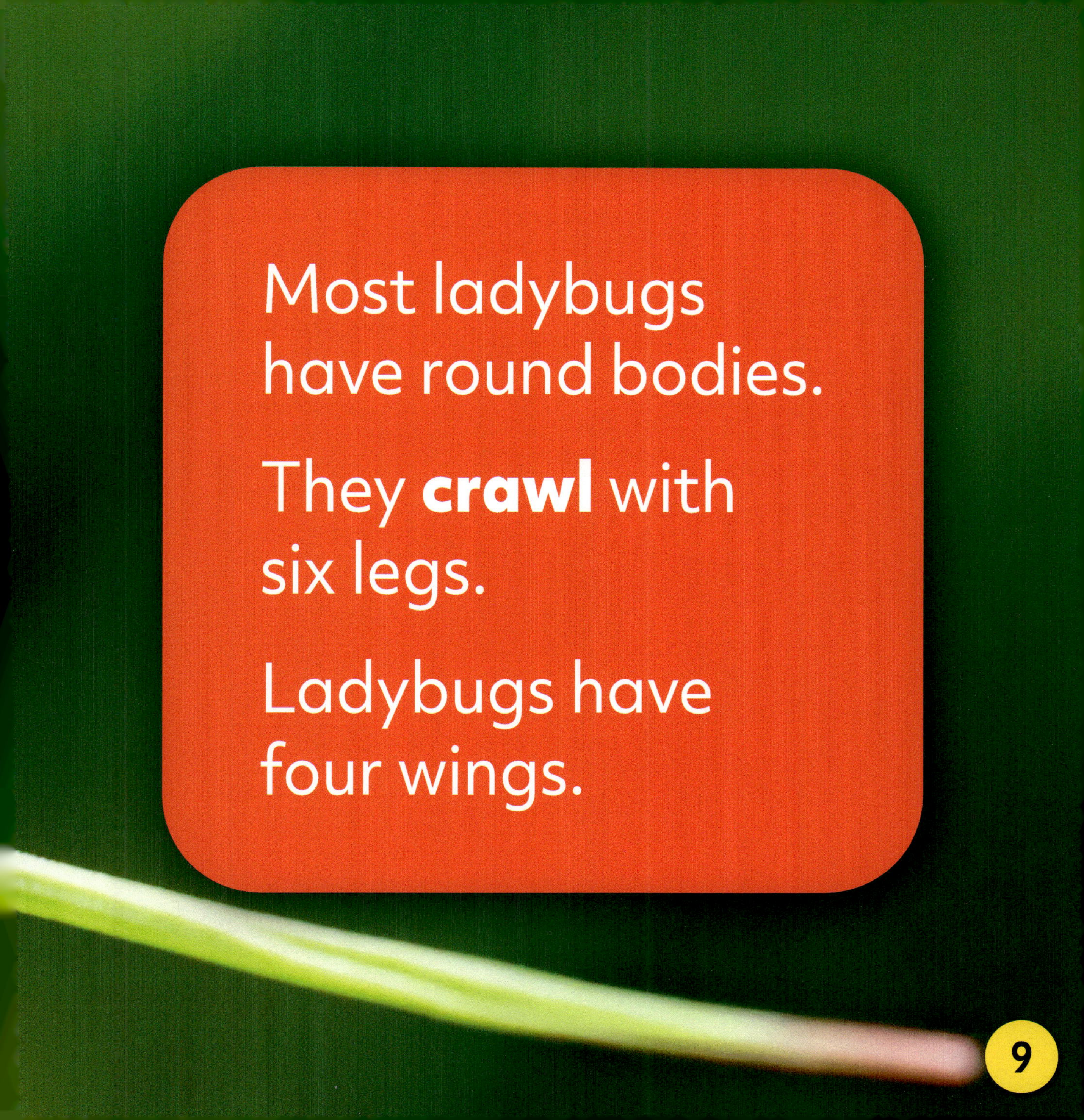

Most ladybugs have round bodies.

They **crawl** with six legs.

Ladybugs have four wings.

A ladybug uses two wings to fly.

When it lands, its flying wings fold under hard wings.

The hard wings are colorful.

Hard wing
Flying wing

Some ladybugs are red.

Others are yellow or orange.

The colors keep ladybugs safe.

They **warn** other animals that the bugs taste bad.

Most ladybugs eat other bugs.

Farmers like ladybugs because they eat bugs that hurt **crops**.

A crop

Ladybugs eat in the summer and fall.

Munch, munch!

They can eat more than 50 bugs a day.

This helps them get ready for winter.

When it gets cold, ladybugs come together.

They pile up in large groups.

Then, they sleep.

In the spring, the ladybugs wake up.

They warm up in the sun.

Then, they fly away.

Ladybugs can live two to three years.

A Ladybug's Life

Say pupa like PYOO-puh

Say larva like LAHR-vuh

Glossary

bright having a very strong color

crawl to move with the body close to the ground

crops plants that are grown for food

warn to tell about trouble

Index

Read More

Carr, Aaron. *Ladybugs (Exploring Life Cycles).* New York: AV2, 2021.

Neuenfeldt, Elizabeth. *Ladybug (Animal Life Cycles).* Minneapolis: Bellwether, 2021.

Learn More Online

1. Go to **www.factsurfer.com** or scan the QR code below.
2. Enter "**Ladybug Bug**" into the search box.
3. Click on the cover of this book to see a list of websites.

About the Author

Martha London lives in Minnesota with her cat. She has written more than 100 books for young readers!